AF588487

GEOLOGY ROCKS!

EXAMINE MINERALS

CHRISTINE PETERSEN

An Imprint of Abdo Publishing
abdobooks.com

ABDOBOOKS.COM

Published by Abdo Publishing, a division of ABDO, PO Box 398166, Minneapolis, Minnesota 55439.

Checkerboard Library™ is a trademark and logo of Abdo Publishing.

Printed in the United States of America, North Mankato, Minnesota
102019
012020

Design: Emily O'Malley, Mighty Media, Inc.
Production: Mighty Media, Inc.
Editor: Jessica Rusick
Cover Photograph: Shutterstock Images
Interior Photographs: Abdo Publishing, p. 9; Getty Images/iStockphoto, p. 20; Javier Trueba/MSF/SCIENCE SOURCE, p. 29; Mighty Media, Inc., pp. 10, 11; Ra'ike/Wikimedia Commons, p. 19; Shutterstock Images, pp. 5, 7, 13, 14, 15, 16 (left and right), 17, 18, 21, 23 (left, middle, right), 24, 25, 27

Library of Congress Control Number: 2019943344

Publisher's Cataloging-in-Publication Data

Names: Petersen, Christine, author.
Title: Examine minerals / by Christine Petersen
Description: Minneapolis, Minnesota : Abdo Publishing, 2020 | Series: Geology rocks! | Includes online resources and index.
Identifiers: ISBN 9781532191725 (lib. bdg.) | ISBN 9781532178450 (ebook)
Subjects: LCSH: Geology--Juvenile literature. | Minerals--Juvenile literature. | Rock-forming minerals--Juvenile literature. | Mineralogy--Juvenile literature. | Rocks--Classification--Juvenile literature.
Classification: DDC 549--dc23

CONTENTS

MINERALS ARE **EVERYWHERE!**

Anne and her family were enjoying their summer vacation. They took a trip to the beach. There, the sand sparkled as it caught the sunlight.

The next day, Anne's family drove inland. The highway followed a river that flowed past green farm fields. The family set up camp in a valley alongside the river. Mountain peaks towered high above them.

The beach, farm field, and mountain valley may seem very different from one another. But these places have something in common. Much of the beach sand is made up of tiny pieces of minerals. The farm field soil is also full of minerals. Even the mountains are made from many minerals packed together.

All minerals have four things in common. Minerals are made in nature, and they are usually **inorganic**. Minerals are also solid substances that have a specific chemical makeup. Finally, minerals form units called crystals. These features will help you recognize the minerals around you!

Earth is covered in rocks made of minerals. ▶

ATOMS AND **ELEMENTS**

Everything on Earth is made of tiny building blocks called atoms. Bonds hold atoms together. In minerals, atoms bond together in repeating patterns. This forms crystals, which grow larger as more atoms join the pattern.

The pattern of atoms gives a mineral crystal its shape. Crystals of the same mineral have the same shape. Different minerals have different shapes.

Elements are made up of atoms of only one kind. Scientists have discovered more than 100 different elements. Atoms of those elements bond in different patterns and combinations. Together, they produce more than 4,000 known minerals.

Most minerals contain two or more elements. The mineral gypsum is probably in your home in drywall or cement. It contains the elements calcium, sulfur, oxygen, and hydrogen. Tourmaline is a mineral that is often used in jewelry. It can include more than seven different elements.

A few minerals contain only one element. Sulfur is made of the element sulfur. Diamond also contains just one element. It is made from carbon.

PERIODIC TABLE OF ELEMENTS

Scientists arrange chemical elements on a chart called the periodic table. Elements are organized into columns, rows, and colors by features they share. Each element has a chemical symbol made up of at least one letter. Some are easy to figure out, while others are not. For example, O stands for oxygen. Pb stands for lead. The symbol comes from *plumbum*, the Latin word for "lead." When scientists discover new elements, they add the elements to the table. Each element receives a name and symbol.

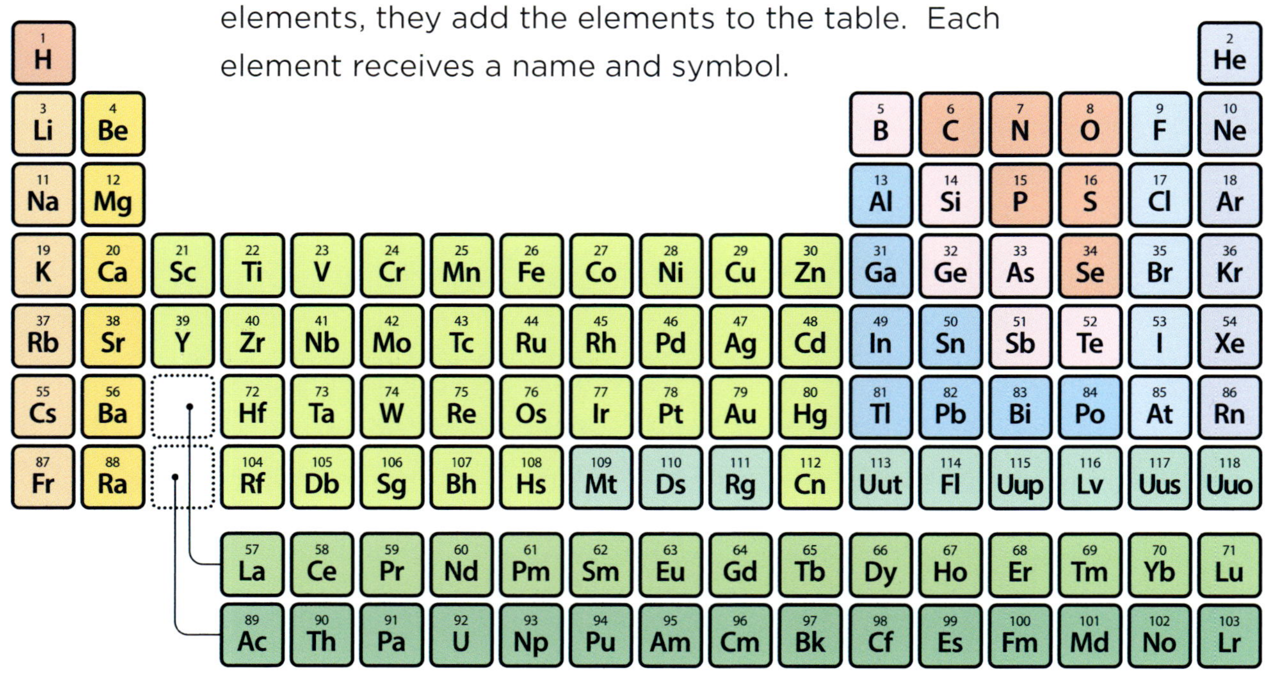

CRYSTAL SYSTEMS

A mineral crystal's shape is based on how its atoms are arranged. Scientists have fit these shapes into seven groups called crystal systems. These are isometric, hexagonal, trigonal, tetragonal, orthorhombic, monoclinic, and triclinic.

The simplest isometric crystals form cubes. The mineral pyrite is isometric. Pyrite is commonly called fool's gold. It can be mistaken for real gold.

Hexagonal crystals may have 6 or 12 sides. They can look like columns or double pyramids. Graphite is a hexagonal mineral. It is used in pencils.

Trigonal crystals may have six **rhombus**-shaped faces. This system is sometimes considered part of the hexagonal system. Calcite is a trigonal mineral. It is one of the most common minerals on Earth.

Tetragonal crystals may have four rectangular sides. Squares form the top and the bottom. Chalcopyrite is a tetragonal mineral. It is a major source of the metal copper.

Some orthorhombic crystals look like two pyramids stuck together. The orthorhombic mineral topaz is a popular gem used in jewelry.

The simplest monoclinic crystals consist of four rectangles and two **rhombuses**. People have long used the monoclinic mineral jadeite for sculptures and jewelry.

Triclinic crystals often look crooked. Their sides have different shapes and do not meet at right angles. Some feldspar minerals are triclinic. Feldspar minerals make up a large part of Earth's crust.

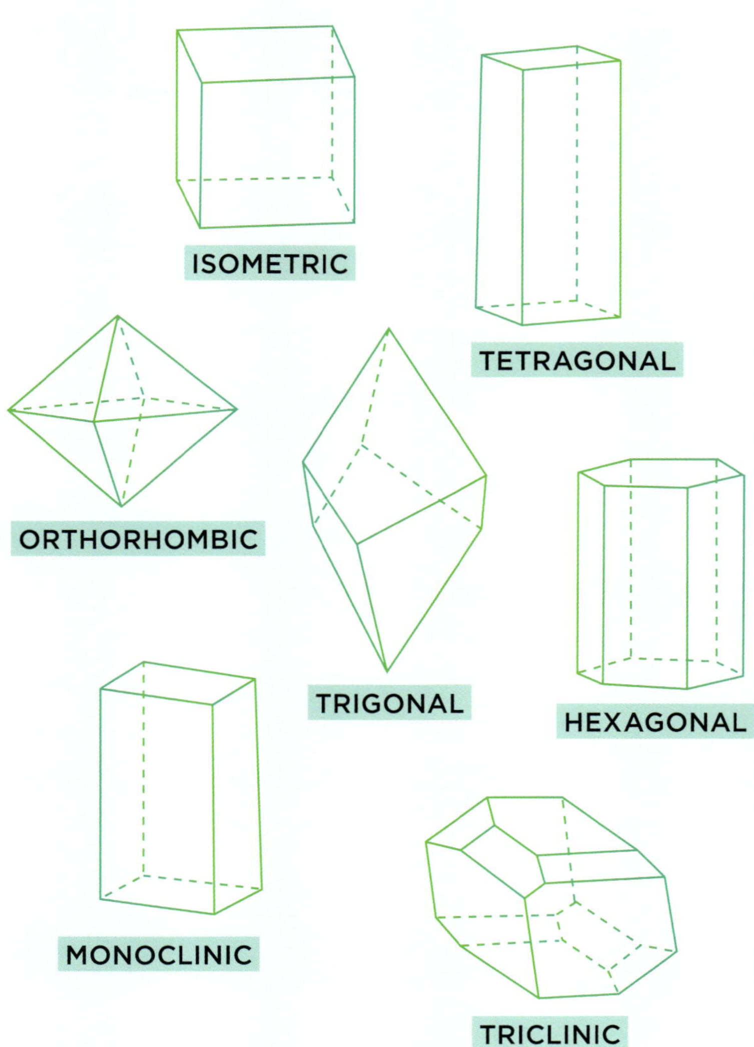

TRY THIS AT HOME: MAKE ROCK CANDY!

WHAT YOU'LL NEED

- cotton string
- scissors
- glass jar
- pencil
- paper clip
- 1 cup water
- medium saucepan
- 3½ cups sugar
- spoon
- food coloring

WHAT YOU'LL DO

1. Cut a piece of cotton string about the height of the jar. Tie one end to the middle of the pencil. Tie the other end to the paper clip.

2. Lay the pencil across the top of the jar. The string should hang about two-thirds of the way down inside the jar.

3. Pour the water into the saucepan. Have an adult help you bring the water to a boil on the stove.

4. Slowly add sugar to the water, stirring until it **dissolves**. This will form a syrup.

5. Stir in three drops of food coloring.

6. Turn off the stove. Let the syrup sit for 30 minutes.

7. Pour the syrup into the jar.

8. Set the jar in a safe place. Crystals will form over the next few days.

9. Pull out the crystal-covered string and take a close look. Sugar is not a mineral, but its crystals do fit into one of the crystal systems. Can you tell which one? The answer is upside down on the next page!

What crystal system do sugar crystals belong to?

They belong to the monoclinic crystal system.

MINERAL FORMATION

Many minerals grow from liquids, including **magma.** Magma bubbles up from deep below Earth's surface. It may stop before reaching the surface or erupt from volcanoes as lava. As the magma cools and hardens, mineral crystals grow. These crystals harden into **igneous rock**.

Minerals also form as wind, rain, and ice wear down igneous rock. Little bits of rock break off, wash away, and gather in a new place. These sediments are made of various minerals. As they pile up, pressure increases and sedimentary rock forms.

When rocks are under great heat and pressure, new minerals form. Deep underground, those forces turn the rock's old minerals into new ones. These changes also create **metamorphic rock**.

Minerals also grow when water freezes or dries up. That means ice is a mineral when it occurs naturally. Its crystals grow when water freezes. Halite crystals form when oceans or salty lakes dry up. Halite is the mineral

Granite forms as magma cools slowly underground. It contains large, colorful crystals of the minerals quartz and feldspar.

name for salt. It can be mined from underground. Halite is used to melt ice on winter roads and even to make homemade ice cream!

THE ROCK CYCLE

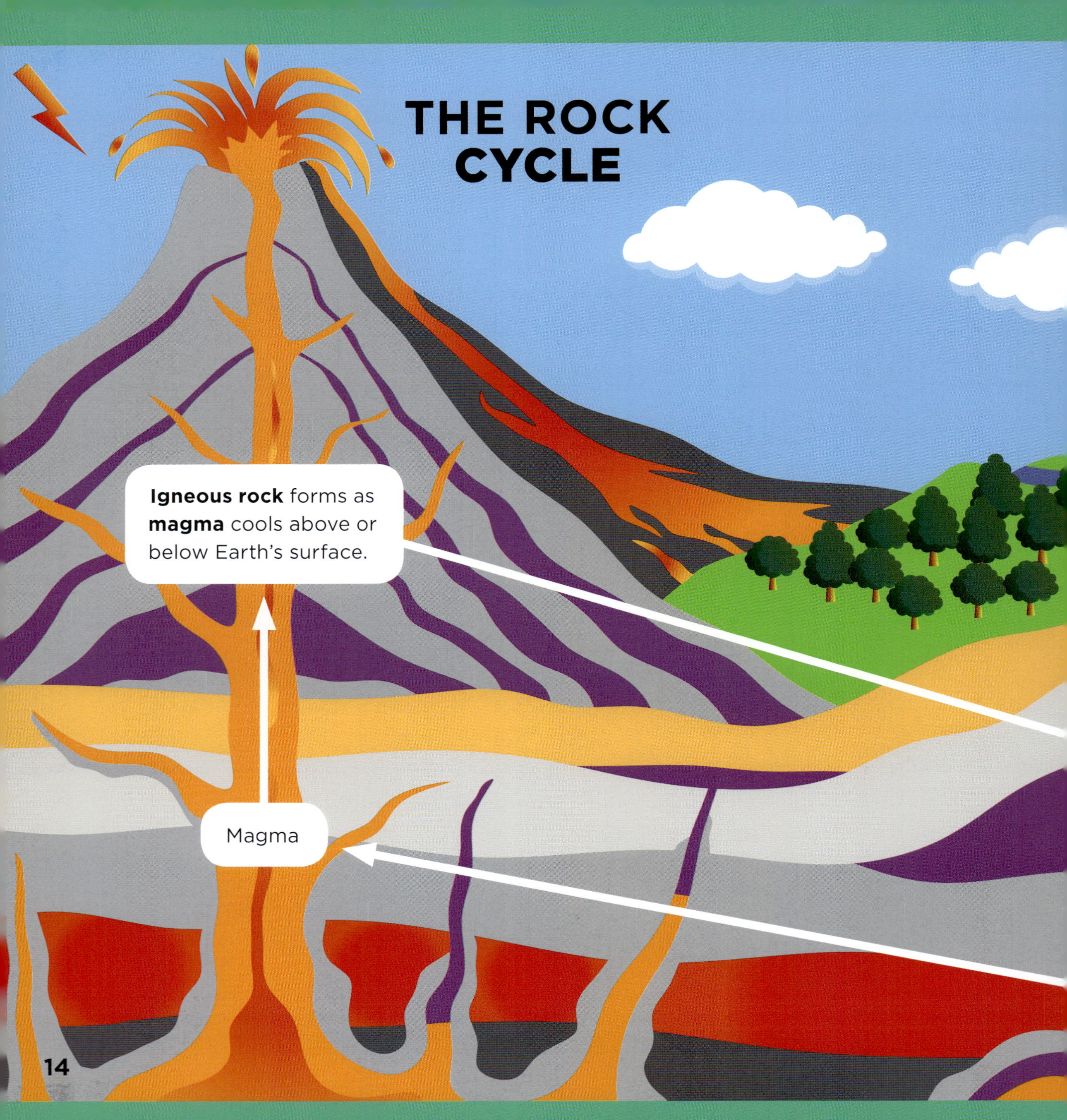

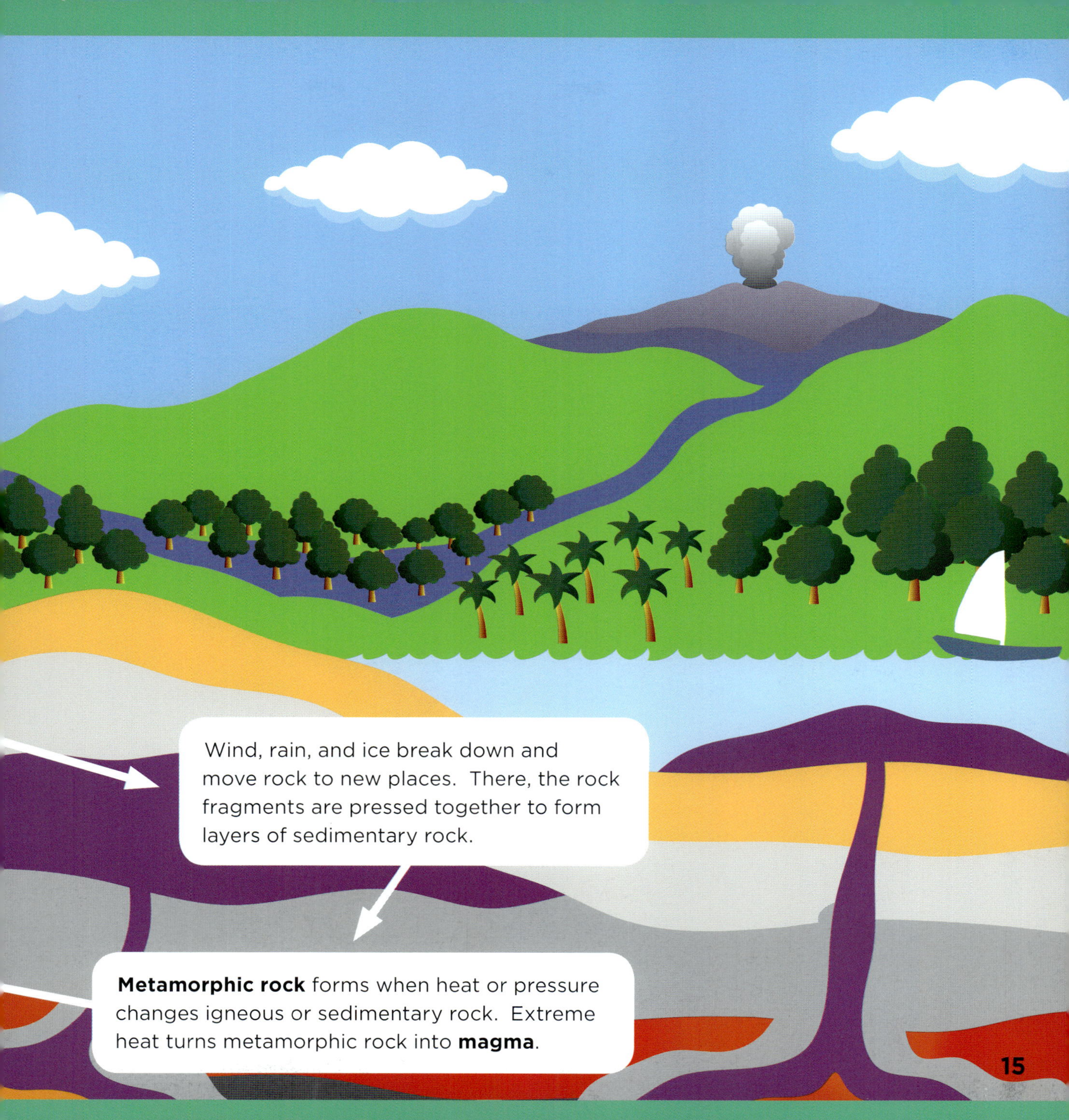
Wind, rain, and ice break down and move rock to new places. There, the rock fragments are pressed together to form layers of sedimentary rock.
Metamorphic rock forms when heat or pressure changes igneous or sedimentary rock. Extreme heat turns metamorphic rock into **magma**.

IDENTIFYING
BY COLOR

It can be hard to identify minerals just by looking at them. So, scientists rely on different physical properties and tests to tell one mineral from another.

Color is one of the most obvious properties of an object. Some minerals occur in only one color. Pure sulfur is always yellow and malachite can only be green. So, color helps identify these minerals.

However, many minerals are found in a variety of colors. The mineral corundum is a good example. Pure corundum crystals are colorless. Rare crystals of corundum may be

Sulfur

Malachite

The mineral azurite is always blue.

bright blue or red. These are the gemstones known as sapphires and rubies. Corundum may also be pink, yellow, or gray.

LUSTER
AND STREAK

Another visual way to identify minerals is luster. Luster describes the way minerals reflect light. Minerals such as pyrite shine like metal. Others are described as glassy, oily, pearly, or dull.

The mineral galena has a metallic luster.

Pyrite (*left*) leaves a greenish-black streak. Rhodochrosite's streak (*right*) is always white.

A streak color test is another way to tell minerals apart. Take a slightly rough **porcelain** tile and scratch the mineral along it. In some cases, the rubbing will leave behind a powdered line of color on the tile.

No matter their color, samples of the same mineral have the same streak color. For example, corundum has a white streak no matter what color its crystals are.

BREAKING MINERALS

Scientists also study how a mineral breaks to help identify it. Some minerals are softer than others. Diamond is one of the hardest natural substances on Earth. It forms deep beneath Earth's crust. There, heat and pressure push its atoms close together. The atoms bond tightly to one another. Yet even the hardest minerals can break.

A smooth break is called cleavage. Minerals that cleave break into pieces with flat surfaces. They also break in certain directions. Diamond breaks in four directions to make pyramid shapes. Halite breaks into cubes. Mica has perfect cleavage. It breaks into thin layers like sheets of paper.

Some crystals do not break into pieces with flat,

Mica

Citrine is a form of quartz. When it breaks, its surfaces are not flat.

even surfaces. Instead, they fracture. Quartz is a mineral that does this. It is a very hard mineral that is common in Earth's crust. When quartz breaks, it has smooth, curved surfaces like a shell. Other minerals fracture in different ways. They may break into splintery fragments or become hackly, or jagged.

HARDNESS AND DENSITY

Hardness is yet another way to identify a mineral. In the early 1800s, German scientist Friedrich Mohs created a scale of mineral hardness. The scale has values from one to ten. One is the softest and ten is the hardest.

Mohs chose one mineral to represent each value on the scale. Hardness is determined by whether or not a mineral can scratch one of these ten minerals. A harder mineral will scratch a softer mineral.

A few everyday objects are also used on the scale. Some minerals are soft enough to be scratched by a copper penny. Their hardness is lower than 3.2. Other minerals are too hard to be scratched by a pocketknife. Their hardness is higher than 5.1.

Scientists also use specific gravity to identify minerals. Specific gravity compares a mineral's weight against the weight of the same volume of water. For example, graphite has a specific gravity of 2.2. It is more than twice as heavy as water. Diamond has a specific gravity of 3.5. Gold and platinum are very heavy. They can have a specific gravity higher than 19!

MOHS HARDNESS SCALE

Mohs Hardness	Mineral	Hardness of Other Materials
1	talc	
2	gypsum	2.2 fingernail
3	calcite	3.2 copper penny
4	fluorite	
5	apatite	5.1 pocketknife
6	orthoclase	5.5 glass plate
7	quartz	6.5 steel needle
8	topaz	7.0 streak plate
9	corundum	
10	diamond	

MAGICAL **MINERALS?**

A few minerals are easy to identify because they have special features. Some minerals that contain iron have magnetic properties. Magnetite has this feature. It is strongly attracted to magnets. Magnets attract iron, nickel, steel, and other objects. Lodestone, a form of magnetite, can even act as a magnet.

Another special feature some minerals have is **fluorescence**. Fluorite crystals are a good example of this. In normal white light, they can appear yellow,

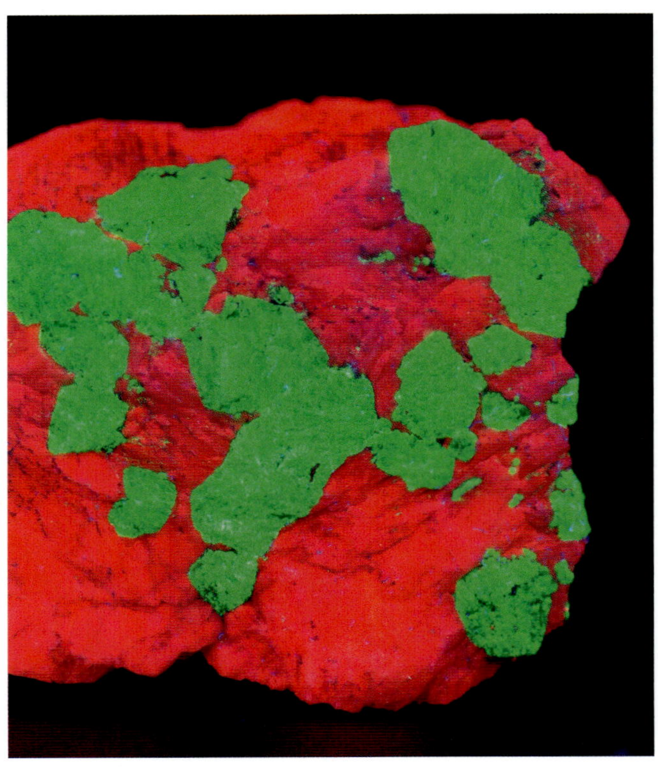

The minerals calcite (*red*) and willemite (*green*) show fluorescence.

The mineral bismuth is repelled from magnetic fields. This is due to how its electrons spin.

orange, black, pink, brown, green, or purple. Yet under special **ultraviolet** light, fluorite crystals usually glow blue or green!

Fluorescence is not always useful for mineral identification. This is because the same minerals can fluoresce different colors. Calcite minerals often have red fluorescence. But they can also fluoresce in a rainbow of other colors, including blue and orange.

PUTTING MINERALS **TO WORK**

People have used minerals for thousands of years. Today, we can't live without them. Take a quick look around your house. You might be surprised by how many objects contain minerals.

The coins in your pocket are made from minerals such as copper. Some toothpastes and drinking water contain fluoride, which helps reduce tooth decay. Fluoride contains a substance that comes from the mineral fluorite.

Quartz can be melted down to make window glass. Test tubes and telescope mirrors are also made from quartz crystals. And, quartz crystals are used in watches. The watch's battery sends an electric current across the quartz crystal. This causes the crystal to vibrate a certain number of times per second. The watch uses the vibrations to help measure time.

Metals can also be minerals. Iron is a strong metal. It can be made into steel and used to make buildings and bridges. Water pipes are often made from copper, which is less likely than other materials to rust.

There is apatite in your teeth and fluoride in your toothpaste. They help keep your teeth strong and healthy.

There are even minerals inside living things. Apatite is a mineral found in your body. It forms part of your teeth and bones. Apatite represents a five on the Mohs hardness scale.

CHANGE HAPPENS

Earth is a place where things are always changing. Minerals are forming and breaking down all around us. They are even growing deep underground in caves.

Over millions of years, crystals grew in Mexico's Cave of Crystals. Minerals from the cave **dissolved** in water. Over time, the solution cooled. This formed gigantic gypsum crystals up to 39 feet (12 m) long! Today, they are some of the largest natural crystals ever found.

Minerals are found everywhere. Mining them from the ground can be damaging to the planet. It can cause air and water pollution.

So, we must use minerals wisely. That way, they will continue to improve our lives long into the future. The next time you visit the beach or the mountains, remember all the minerals that surround you!

The Cave of Crystals was discovered in 2000 in a flooded mine. Scientists studied the crystals after the water was pumped out. In 2017, the cave flooded again. ▶

GLOSSARY

dissolve—to pass into solution or become liquid.

fluoresce—to give off visible light when exposed to a different type of radiation, such as ultraviolet light. Substances that fluoresce have fluorescence.

igneous rock—rock formed from melted rock material that cooled and became solid.

inorganic—being or made of matter other than plant or animal.

magma—melted rock beneath Earth's surface.

metamorphic rock—rock formed from other rock that was changed by heat and pressure.

porcelain—a hard, fine-grained, transparent and white material.

rhombus—a shape with four equal sides and sometimes no right angles.

ultraviolet—a type of light that cannot be seen with the human eye.

SAYING IT

chalcopyrite—kal-kuh-PEYE-rite

fluorescence—flu-REH-suhns

Friedrich Mohs—FREE-drihk MOHS

igneous—IHG-nee-uhs

malachite—MA-luh-kite

metamorphic—meh-tuh-MAWR-fihk

orthorhombic—awr-thuh-RAHM-bihk

porcelain—PAWR-suh-luhn

rhombus—RAHM-buhs

sapphire—SA-fire

ONLINE RESOURCES

To learn more about minerals, please visit **abdobooklinks.com** or scan this QR code. These links are routinely monitored and updated to provide the most current information available.

INDEX